Bitterest Truth

Poems by
Dennis J. Jackson

Bitterest Truth
Dennis J. Jackson

Copyright (c) 2023 by Dennis J. Jackson.

All rights reserved. No part of this book may be used or reproduced in any form whatsoever without written permission except in the case of brief quotations in critical articles or reviews.

Printed in the United States of America.

For more information, or to book an event, contact:
Words to Ponder Publishing Company, LLC
25 Garden Park Drive
Chagrin Falls, Ohio 44022
https://www.wordstoponderpublishing.com

Book design by
Words to Ponder Publishing Company, LLC

ISBN 978-1-941328-12-5
Ebook 978-1-941328-13-2

First Edition : March 2024
10 9 8 7 6 5 4 3 2 1

To The Veterans:

    I wholeheartedly dedicate this entire work to every individual who has valiantly worn the uniform of the United States Military. Your unwavering commitment, dedication, and sacrifice serve as an enduring inspiration. Each one of you is not just a hero, but a beacon of courage and honor.

                                                Dennis

Table of Contents

Bitterest truth	ii
Bitterest truth	iii
To the Veterans	01
a soldier's sacrifice.	06
bitterest truth	07
everyone bleeds green.	08
disposable heroes.	09
8th day	11
22 each day	12
a dream	15
a mother's love	17
a soldier's sacrifice.	20
a Veteran's hope for healing	21
as I saw you leaving	23
at just two	24
bacon.	26
battle scars	27
bourne of two fathers.	29
brothers of the risers	31
by grace alone	32
dad	34
darkest nights	35
dust in the wind	36
each day	38
every now and then	39
every soldier's battle	40
dear dad	42
five senses	43
forever in time	45
four seasons	46

friendship tree	47
gentle winds	48
ghost warriors from the sky	49
Goshen	50
grandpa	51
He provides	53
he was 22.	55
hospital vets.	58
if love's a flame.	60
if you weren't there	62
in state	63
in the darkness	64
indifference	65
investments	66
is it today	67
I've gotta go.	70
Jesus, mom, and me	71
jump wings	72
just a regular Veteran	73
just wanted	74
knights	76
life notes	78
lonely holidays	80
Memorial Day	82
mental health.	84
military honors	86
my flag ii	88
my grandpa	89
my spangled banner	90
nights of roaring silence.	91

oh, so long	93
on my back	95
our military codes	97
part of me is gone	99
passage of time	100
picture of perfection	101
real friends	102
selfless sacrifice	104
serenity	106
service animals	107
service connected	109
set free	110
set your pack down	112
she or he died	113
sisters and brothers	114
skeletons and shadows	117
someone	119
stand up	121
stranger	122
struck by lightning	123
suffer the children	124
tears for the fallen	126
tears	127
the American soldier	128
stages	130
to each it's own mile	132
thirteen	134
souls lament	136
the picture window	138
true soldiers in green	141

the earth's been hushed	143
the sharpest blade	145
whispers in the silence	146
tiny pieces	147
the pack on your back	149
thankful	151
treat us gently	152
springtime springs	154
while everyone sleeps	155
the old man in the fall	156
the silent march II	157
when we sign	158
every soldier's battle	159
to what a man's soul	161
the Veteran effect	163
the guard	164
the dotted line	166
the lone veteran	167

a soldier's sacrifice

my military service
honor to behold
passing through generations
spanning days of old
when our passing arrives
angels sent from above
surrounded by family
all of those we love
special places in heaven
for peacekeepers, one and all
an understanding father's word
when each of us finally falls
welcome home my servant
your battles are surely won
your life is full of strife
it is finally done
willingness to serve others
outside than yourselves
like capturing your hearts
and keeping them on shelves
well done good and faithful servant
come in and put down your pack
the good Lord has this covered
the angels have your back

bitterest truth

the bitterest truth is better than
    the sweetest lies
no worries concerning alibis
speak the truth for all concerned
both heart and mind are never burned
you can change the world
one soul at a time
just jotting down
a heartfelt rhyme
share your love
with those you know
take the bitterness
and let it go
each new day
a chance to dance
to spark the fires
of a new romance
speak the truth
every word you say
never a penalty
will you have to pay
share God's love
in all you do
his amazing love
will carry you through

everybody bleeds green

every day we realize
when a uniform is seen
the meaning of the phrase
that we all bleed green
a soldier or sailor
airman or marine
acting in their best behavior
your heart starts bleeding green
at every family gathering
there's always someone gone
your heart is bleeding green
everything just feels wrong
on public transportation
when every seat is gone
the person giving up their seat
green blood is flowing strong
at every sporting event
during America's greatest song
we all feel that heart swell
green blood is flowing strong
when a Veteran helps someone
expecting nothing in return
green blood flowing softly
dreams of respect in return
so, you see so clearly
in almost every scene
if you keep an eye out
you see, we all bleed green

disposable heroes

from early on
to present day
every soldier
was a throwaway
do the dirty work
take the blame
hide the pain
inside like shame
do the things
no one will
until your body
is cold and still
freedom is
a fleeting call
for the few
who stand so tall
screaming nights
death in sand
some will go
by their own hand
so people can spit
disrespect their gift
call them losers
cause a rift
freedom costs
is truly dear

for all of those
standing near
like paper cups
or plastic bags
the tired old Veteran
wearing tattered rags
he has his pride
his private hell
and too many demons
he knows so well
a simple tear
streaks his face
saluting his flag
and amazing grace
society tossed him
like useless trash
giving stupid athletes
way too much cash
a hero dies
for minimum wage
the entire country
should be full of rage
disposable heroes
another every day
every one of us
wouldn't have it any other way

8th day

we all know the story
how God created heaven and earth
but then he created something
of amazing value and worth
as time passed in eons
the Lord gave us the greatest gift of all
only certain family's get lucky enough
to have such a sister-in-law
she opens her heart to each of us
makes us feel special at every turn
the special gifts she brings along
in no measure, can we return
she is definitely irreplaceable
gifts always chosen from the heart
there truly is never a doubt
she's the right one from the start
on the eighth day, God got started
on a special project for the world, to see
a special mold for sister-in-laws
like the one you are to me!
love ya, Jen

22 each day

as a small boy
each day would play
never played navy
was airborne all the way
knew how to hide
aways unseen
stick for our weapons
belts used as a sling
then we grew up
finally finished school
working two jobs
really just wasn't cool
signed our names
on that dotted line
leave the old world
truly far behind
took to the training
living the dream
quickly becoming
a killing machine
protecting civilians
in foreign lands
oppressive heat
and blowing sand
in jungle heat
torrential rains

underground cages
to destroy our brains
arriving home to
insults and rejection
causing us each
deep inner reflection
we served our nation
was proud of our flag
yet now it feels like life
is such a drag
doctors say words
acting in a play
do not know why
we feel this way
have never dodged bullets
jumped from the sky
put their life on the line
been ready to die
they've never run
seventeen miles fully loaded
yet sit in their chairs
all Veterans being goaded
make vets feel worse every day
fake doctors without care
will it happen today
is this your dare

we come to you for treatment
compassionate care too
instead you dare us each
to be one of today's 22
you should check your ego
when you enter those doors
unless you believe
the number should be more
when we head for the door
regardless of when
the doctors shouldn't stab us
with their poison pen
remember each morning
as you begin to pray
a Veteran's sacrifice
has made it that way
we never asked you
to carry our packs
just keep your knives
out of our backs
when we ask for help
make sure that it's true
make sure we don't become
one of today's 22
help bring an end to Veteran suicide

a dream

finding words
perfectly read
fulfilled dreams
nightmare's shed
a calming voice
in every season
things make sense
every rhyme and reason
a state of mind
before unknown
a path of light
finally, been shown
to best describe
what I can see
the entire world
as a tiny pea
or the world
despite its pace
isn't always
a lonely place
eyes before
had never shared
the meaning of love
that's being shared
two hearts beat
as if one

who knew love
could be this fun
it's a peaceful
heart-warmed place, sharing love
in our shared space
i've never known
peace this way
I pray you never go away

a mother's love

a mother's love
is a special thing
it fills her heart
and makes her sing
a mother's love
grows each day
showing in every
word they say
a mother's love
is always shown
to every child
it's well known
a mother's love
is always there
helping each child
grow and prepare
a mother's love
like a good luck charm
keeping us safe
from danger or harm
a mother's love
is like Miracle Grow (R)
just a little each day
and miracles show
a mother's love
as most will see

create the roots
of the greatest tree
a mother's love
always giving the best
preparing us
for the final test
when our mother's
must pass beyond
they've made sure
we're big and strong
our hearts are ready
our heads are sure
for this life's struggles
we each must endure
a mother's love
on this life's plane
a thriving force
none can explain
it goes beyond
this lonely earth
dating back to
the earth's birth
God created
a mother's love
to help us all learn
to rise above

to carry on
what we were taught
to show each day
it wasn't for naught

to all who have lost their moms and realize every day what they were taught and how valuable each lesson was. I love you, Mom.

a soldier's sacrifice

my military service
honor to behold
passing through generations
spanning days of old
when our passing arrives
angels sent from above
surrounded by family
all of those we love
special places in heaven
for peacekeepers, one and all
an understanding father's word
when each of us finally falls
welcome home, my servant
your battles are surely won
your life full of strife
it is finally done
willingness to serve another
other than yourselves
like capturing your hearts
and keeping them on shelves
well done good and faithful servant
come in and put down your pack
the good Lord has this covered
the angels have your back

a Veteran's hope for healing

every single passing day
another Veteran needs healing
damaged by the facts of war
some better ways of dealing
ragged are the tragedies
being dealt to our soldiers
too many checking out
before they get much older
no one seems to understand
the battle rages on for real
twenty-four/seven/three sixty five
no escaping what we feel
it's very hard for you to understand
the hell we suffered through
we each gladly do it again
to prevent it from coming to you
every morning, noon, and night
the battle rages on and on
foreign noises, strangers' voices
reminders of comrades that are gone
we gather at the hospital
AA meetings and in bars
for our military brethren
o distance is too far
we value our male Veterans
females are special as well
we all served with honor
in different types of hell

some chose to self-medicate
while others choose to abuse pills
thinking it's the easy answer
to all of their daily ills
we each must find the answer
in exercise, meditation, and diet
learning to help each other
just keeping those demons quiet

as I saw you leaving

today wasn't a good day
as we silently watched you leaving
left behind are all of those with love
forever left here grieving
the whole world could never know
the magic of your smile
or the way you'd lift our hearts
by visiting for a while
they could never see
the love inside your heart
and your connections with family
that really set you apart
no more joyous laughter
no more hugs to be received
the Lord has called you home
with tears upon our sleeves
we only know for certain
for the good Lord has made it so
we will see you again
when it's our time to go

at just two

life is just
a prison cell
oft we stray
know thee well
fire-haired child
third in line
strange friendship
heavenly design
the main thing
for people to know
is what involved
a tiny hero
at what was then
less than two
he understood then
what heroes do
a bedroom fire
crib in flames
putting himself
between me and flames
I know few people
can make that stand
kicked fire's ass
and saved a man
I saw his scars
like superman's shield
fought that battle
and with God's help healed

all my life
tried to repay
the one thing
words cannot say
may his spirit
finally, be free
for the life he saved
and gave to me

bacon

nights without bacon
never seeing the end
I think there's a diner
up around the bend
three strips on a plate
never will suffice
it's gotta be
six strips to paradise
think it's an apparition
in this subdivision
do-do-do
grill my bacon some more
in the shuffling madness
she's got bacon breath
like a sandwich winner
heartache away from death
in the lights of the city
in a diner by the bay-ay
they'll make Canadian bacon
every single which way-ay
oh, ooh whoa!
ooh whoa!
bacon made my way
guess I'm hooked on bacon

battle scars

we each carry
our battle scars
individualized
like plates on cars
some are hidden
for few to see
like the bark
of a mighty tree
others are vivid
I guess you'd say
marks of bravery on display
some are colorful
the message plain
a silent reminder
that keeps us sane
each a reminder
of battles won
pain defeated
by tattoo gun
body art
silent tributes
every warrior
is resolute
friends lost
and family dear
on display

for all to hear
the breaking heart
the lonely cry
the soldier bidding
a final goodbye
each scar
a testament
each soul
is heaven sent

bourne of two fathers

only certain gentleman
will understand this part
bourne of two fathers
in this telltale heart
I was bourne to city folk
their hectic daily ways
longed to be in the woods
every single day
no man would take the time
in spite of whispered screams
it took another forty years
to realize my dream
a gentleman and his father
saw inside my dreams
took me into the woods
I was ready to live the dream
I knew the code of silence
steps among the mighty pine
if I'm not following your steps
you will be following mine
the taking of an animal
for sustenance, not for game
occasional rack for mounting
it all tastes just the same
leave for other creatures
that what God intends

all of the bits and pieces
all of the loose ends
never waste a single thing
appreciate every single meal
do what God intends
don't lie, or cheat, or steal
another father found
at the age of forty-eight
a different type of person
would slowly integrate
freedom in the forest
close with animals
as they abound
just a man and his thoughts
with no one else around
gentlemen built my spirit
I knew I had the stuff
just do what my God would ask me
be ready to call a bluff
using things in nature
balance for everyday
it's why he created the animals
and humans on the 6th day

brothers of the risers

we are the product of Ben Franklin
a genius ahead of his time
he even helped to conceptualize
the modern-day static line
many troops around our world
have utilized the static line
in America, August 16th of '44
we joined that airborne line
from Sicily to Salerno
to saint mere Eglise and Normandy
practiced more at Holland
always avoiding the trees
all named for past conflicts
where our brothers fought and fell
is why we wear this emblem
a badge of honor and pride as well
every airborne unit has been
as you find through the test of fire
America's greatest patriots proven
by their strong spirit and desire
from the 1940s
repelling the German brigades
paratroopers stopped their invasion
with chutes, tanks, guns, and grenades
they took control in Korea
Vietnam, Iraq, and Afghanistan
proving to the world what's needed
is another tough jumping man
silent death from above
land soft then kill quiet
chewing up the enemy
that is the airborne diet

by grace alone

it was by grace alone
that I am here drawing breath
it is by the Lord's hand
that I have cheated death
it is by grace alone
that I've passed death's door
it's only a miracle
that God's not keeping score
it is by grace alone
that I share time with others
that I was blessed with family
parents, sister, brothers
it is by grace alone
that my words can help others
sitting alone with my telephone
rhyming hearts with others
it is by grace alone
that air sustains this life
is this soul just tattered
worse than by a knife
it is by grace alone
my lungs can still draw air
that my senses are still sharp
my brain fully aware
it is by grace alone
that I craft these words for others

to gently heal the heartache
of the sisters, dads, and mothers
it is by grace alone
each word falls into place
helping to dry those tears
bring a smile back to your face
it is by grace alone
by God's hand, I write
I just write the words
hoping to help things be alright

dad

Father's Day
is once a year
a single day
we wish him near
a special man
to quite a few
recognized
by what they do
some may not
have it as nice
a life so rough
like living it twice
yet most of us
have a special dad
making every day
feeling extra glad
funny type of humor
always with a joke
sitting on the porch
sipping a Coke (R)
how much I would give
how much I would pay
to have my dad here
this Father's Day

darkest nights

very few people realize
what happens in the depth of night
how drastically things can change
before the morning light
as they slumber safely tucked away
dreams are filled of tomorrow
not a single solitary thought
of those who prevent real sorrow
first responders nationwide
ready for any hour's call
giving back the breath of life
to strangers big and small
firefighters, police, paramedics
trained to save a life
often paying the ultimate price
leaving kids and wife
military reservists are next
ready to help in times of need
responding as quickly as possible
with sometimes amazing speed
active-duty military personnel
front lines against foreign aggression
always ready and willing to do
whatever it takes for suppression
we each value our democracy
freedom to live and love
enjoyed every day
granted by the Lord above
when you see the star-spangled banner
waving in the wind at night
see it as the reminder
you are all protected throughout the night

dust in the wind

hours before sunrise
green ramp in the past
we head ever skyward
we know it won't be our last
we are just dust in the wind
red-yellow streamers
as the clouds break a clear dawn
riding this Globemaster
with our chutes and reserves on
we are just dust in the wind
where are you landing
in a stream, lake, or tree
will you get a good PLF
like the guy in front of me
we are just dust in the wind
we are a band of brothers
yes, it runs in families
from uncle to nephew to brother
we are the best that we can be
we are just dust in the wind
we have a code of honor
deeper than any river can run
and when we have the chance
we go and jump for fun
we're just dust in the wind
we are the country's guardians

the first in any attack
you will never see one of us
cry or turn our back
we're just dust in wind
we gladly pay our tribute
to the good Lord up above
that fills our chutes each time
and fills our hearts with love
for
we
are
just
dust
in
the
wind

each day

each day we need
to keep this in mind
never a truer heart
than a mom's you will find
each day remember
the gifts that she gave
the ones she learned
from the one sent to save
each day remember
the lessons she gave you
the gifts from God
that gets us all through
each day remember
all the ways she carried you
and the ways you need to
help to carry her through
each day remember
like the start of life itself
all the ways each mother
put her needs upon the shelf
each day remember
scripture says to honor
every day of life
mother above all others
each day remember
to praise the Lord above
and thank your lovely mother
for teaching you about love

every now and then

every now and then
like writing with a broken pen
no matter what try to I say
feelings get in the way
it seems so few remain
who clearly make that claim
always speaking the truth
learned so hard in their youth
sitting at their bedside
that Veteran's finished their ride
no family for to comfort
we act as their escort
losing any Veteran friend
a wound that never mends
spilled that damn puzzle again
all those wounds that just don't mend
can't they take all these pieces away

every soldier's battle

every soldier's battle
spanning all of time
one of abandonment
of being left behind
every battle fought
since the start of time
forming lasting bonds
in the worst of times
whether bow and arrow
hatchet or M-16
every single soldier knows
things you've never seen
they never asked for pity
you'll never hear them scream
hidden very deep inside
a very terrifying dream
it may manifest as depression
or another type of madness
every soldier carries with them
a high degree of sadness
a fascination with giving
life for fellow man
best to give your life
making a final stand
every single American
anyone worldwide

the soldier will protect
with their greatest pride
the soldiers battle being
a single matter of life
regardless of those left behind
his family or his wife
every single soldier
battles to see each day
as a reason to continue
as a reason to stay

dear dad

when I was little
your hands seemed so large
so strong and muscular
you could carry me so far
you would work your fingers
to the bone the saying goes
from sunrise to sundown
making sure we had nice clothes
you were a great father
taught us life's lessons oh so well
respect, trust, honesty, truth
each lesson rings true as a bell
each lesson always helps us
each day as we carry on
he taught us important lessons
it's like he's never really gone
each time we speak with honesty
show respect and trust once in a while
our father is looking down from heaven
with a very happy smile
whether son or daughter
he was proud of us either way
sharing any time with us
always made his day
you broke our hearts by leaving
but we knew your work here was done
you did a perfect job
of raising a daughter or son

five senses

in the beginning
God created man
and a woman
we each understand
each sense miraculous
most definitely true
like describing a flower
the color of blue
the scent or taste
of a lover's skin
the amazing chance
for love to begin
the tiny tinkle
of piano keys
just reaching ears
seems to please
various colors
everywhere we see
telling differences
steel and tree
the taunting scent
of farm fresh honey
the dry dirt smell
of paper money
the feel of grass
between your toes

or pepper dust
inside your nose
miracle of life
each breath we take
not one day
should we forsake
share this life
put away misery
brighten each day
of our destiny

forever in time

while life's always a mystery
people find throughout history
when their lives fall apart
how to hold the broken heart
with every tear silently falling
the heart's just barely crawling
while we each try to stand
we hold a heart inside our hands
the simplest words oft shared
proof positive that you cared
the opening of a heart
like decades spent apart
someone filling the empty space
with such an amazing face
helping every word to rhyme
real forever in time

four seasons

when I see the crystals
of a winter's morning
each new day becomes
even more reassuring
with the colorful buds
of every spring day
give everyone notice
better weather on the way
the wonders of summer
flowers and fruit galore
evening campfires
chocolaty s'mores
the blazing spectrum
of God's palette in fall
he never skipped
one color at all
with all this beauty
we need to look around
it is in each other
where beauty is found
find beauty in
those you just meet
and your loved ones
that's your greatest treat
yes, you are beautiful
God's greatest gift
now find this in others
and give them a lift

friendship tree

starting friendships can
most surely be
compared to life
and growing trees
each part starts
quite simply, they say
wear your best smile
each and every day
learn new ways
to communicate
start today
it's never too late
a few kind words
a helping hand
are quickest ways
others understand
be the friend
you wish you had
day by day
you'll be really glad
your wildest dreams
start coming true
by being the friend
you draw towards you
friends are definitely
our greatest treasure
they lift us up
far beyond measure

gentle winds

like the winds of time
gently tinkling wind chimes
the people we know
that seem to come and go
some will linger forever
some will seem like never
forever someone is in our heart
time can never tear us apart
some seem like springtime showers
lasting only a few hours
certain few are etched in stone
never quite leaving us alone
no matter where the winds may blow
no matter where you may go
no matter the years apart
you're still deep inside my heart

ghost warriors from the sky

we all knew where we going
from when we were tiny tykes
we all wanted to emulate
the men like iron mike
all the things they taught us
as we became truly airborne men
all the skills and values
we've carried on since then
we've clung to higher values
stood for what was right
watched over America deep
into the night
we trained in every scenario
so to never be surprised
to always come out victorious
against enemies so despised
iron mike gave us spirit
confidence in ourselves and teams
he helped this little boy
live out his childhood dream
few finally make it
of all those who ever try
be proud to call yourself
ghost warrior from the sky

Goshen

we've all been blessed
our personal Goshen
the Lord's crucifixion
his life's devotion
John the Baptist
preparing the way
the Lamb of God
telling him what to say
every sin washed clean
for just the asking
the Lord is ready
with Goshen basking
sincere heart
maybe bended knee
just open a heart
offered to thee

grandpa

I was a child
no more than 3 or 4
but to my grandpa
I guess I was something more
just a tiny boy
hair all fiery red
so often hurt by
things that other people said
this tall gentleman
so many years apart
after all, he'd suffered
touching a young man's heart
suffering through war
and hardest of times
years of struggles
working in coal mines
just a tiny boy
at the tiny age of four
taught to tie his shoes
one day so much more
instead of just a boy
he saw the real man inside
the man he was building
also serving with pride
that man that he built
helping to build others

sharing his love and grace
with Veteran sister and brother
to see that it started
with a grandpa's love
the gentlest words
from the Lord above
to find inspiration
in those gentle acts
is to share with others
what this world lacks
thank you, grandpa
for tying my shoes
and that special love
that's in all I do

He provides

every day
the Lord provides
to see what's best
no choosing sides
to accept
what we need
we are each
a tiny seed
each day
as we grow
the Lord's truths
we should know
everyone
just the same
never be first
in casting blame
share your gifts
each is rare
with a love
that's always there
give away
your greatest treasures
the next reward
far beyond measure
to see beyond
what others see

an open heart
and bended knee
put no value
on rich man's wares
a greater life
as we all share
a kinder word
like a velvet glove
will ring true
the feeling of love
he put us here
to live as one
from the days
the earth begun
love is meant
for all you see
a better life
will surely be

he was 22

he was a skinny young man,
neither black nor white
loved his family
prayed every night
decided to do
what was right
signed his name
on the dotted line
married his sweetheart
from high school
knew she'd never treat
him like a fool
went off to train
then jump school
keeping his faith
in things divine
twenty-seven months
in the desert sands
worked his way up
the chain of command
sworn to protect
each and every man
never knowing the toll
on his heat baked mind
finally came home
to an empty house

learning the toll
of an unfaithful spouse
a screwed up system
forced to espouse
memories of battles
replaying in his mind
no off button
no quick release
no way to make
the pain ever ceases
heat baked brain
just seeking quick release
the establishment now
is totally blind
twenty-two a day
going for the out
not with a whisper
with a shout
America needs to understand
what this is about
there's a growing number
there's not much time
some use a pistol
brains on the wall
watching in public
as his body falls

no one there
would take his call
maybe then
they'd see the sign
some do it in private
like a roulette wheel
at the tip of a needle
no need to feel
will this one kill me
is God real
why can't they help me
fix my mind
twenty-two a day
can't you understand
this has gotten
way out of hand
we need a solution
a helping hand
before it takes
a friend of mine

hospital vets

if you know a Veteran
or are one already
a few quick hints
to help life stay steady
some really great friends
that makes you laugh
it will seem to cut
the days in half
friends that call
on the telephone
letting you see
you're never alone
friends that share
their problems too
making you feel helpful
when the day is through
if you're quarantined
inside your room
talk to people
FaceTime or Zoom
it is like being there
sharing a laugh
lifting up your heart
loneliness cut in half
the virus causes
difficult visitation

sometimes making this
a lonely vacation
but special people
Veterans one to another
such caring hearts
like sisters and brothers
each is sent
with angel's wings
to inspire one another
with the hope that brings

if love's a flame

if love's a flame
beginning small
very quickly
consuming all
the heat of passion
just tiny embers
sharing times
for each to remember
firsts together
just holding hands
making sure that
hearts understand
sharing time
like rainy days
special memories
to store away
hand in hand
heart in heart
feel the ache
when you're apart
a sandy beach
a warm park bench
a growing desire
difficult to quench
snowy days
stuck inside

adapting when
hearts collide
if love's a flame
a consuming fire
make it last
and truly inspire
make that flame
consume your heart
and warm the days
when you're apart

if you weren't there

if you weren't there
every single day
I would never have
any words to say
if you weren't there
to help me through
I really don't know
what I would do
if you weren't there
to read my words
I wouldn't know
my cries were heard
if you weren't there
to just understand
I might die
by my own hand
if you weren't there
to share my pain
how much of me
would still remain
if you weren't there
to help me through
I might be one of today's 22
if you weren't there
to be a friend
my life of service
would finally end
if you weren't there
what would we do
we need the help
of each of you

in state

lying in state
stars above my head
unable to hear
any words being said
gave up my freedom
went to a foreign land
breathed my final breath
lying in the sand
knocking on heaven's door
not one last breath to take
standing up for judgment
is when I will awake
call us freedom fighters
death from up above
we swore to love our country
and pledged to families our love
lying here is an honor
giving my life for those afar
every stripe running in my veins
in my eyes was every star
as I lie here having passed
for every one of you
the sacrifice was worth it
every word rings so true!

in the darkness

in the complete darkness
of the dark and silent nights
I use you as my lantern
your image my guiding light
in the murky shadows
of the fall or springtime storm
yours is the constant beacon
that essence that keeps me warm
in the rising sunlight
like a compass to lead my way
it is your presence here
that helps me not to stray
in the early evening
as my workday starts to ease
I sit and dream more of you
before you slide into my dreams with ease

indifference

I know not what the cause is
why you see me as some fool
like maybe I'm a halfwit
that never went to school
I have never asked for anything
outside of a chance to share
just to be kicked in the teeth
and shown how much you care
am treated more as a bother
that you wish would go away
this is your final chance
I give up and am walking away
if that's all that am to you
just do as you have done
know the pain that's in my heart
feels like you used a gun

investments

life is such a long road
hard for some to see
how to connect the dots
for the good
things to come to be
simple common gestures
hello, or flash a smile
people start to know you
seeing you once in a while
quick little interface
help to untangle a mess
has the same feeling
of a heart's caress
we must learn each day
to do our part
take time to pray
keep an open heart

is it today

in line of service
celebration of a year
came death for many
my number so near
a life of pain
that only grew
a crippling effect
so few have knew
doctors for years
say such mean things
no rollercoasters
barrels or dangerous things
crazy ass injections
only to aggravate
fueling my anger
frustration and hate
only towards myself
for letting things go
just get through today
you never know
will the good Lord take me
and end my stand
enduring for him
you understand
so many Veterans
lose their way

the pain is too bad
can't handle today
the easy way out
no mess for others
one quick stick
and slowly smother
possibly it's today
sitting in the sun
to say goodbye
with a loaded gun
and yet we remain
trying to be our best
remaining as sheepdogs
so others can rest
we share our hearts
prayers and dreams
so the world won't hear
our muffled screams
for every American
I'd gladly repeat
my days of service
never accept defeat
helping my comrades
one at a time
saving each life
with words that rhyme

if my words inspired
or saved your day
take these words
and give them away

I've gotta go

sitting in the back of this greyhound bus
just trying to forget the days
it seems like a thousand yesterdays
since you left my heart this way
with each and every sundown
my heart just bleeds for you more
the lights flash past the windows
and the tires continue to roar
each mile I get further from you
yet my heart is still stuck in one place
alongside your sensitive heart
and amazingly beautiful face
and yet my heart must go further
for it knows that it cannot remain
for all the pain of yesterdays
we know that love was slain.

Jesus, mom, and me

mother's day quickly approaching
so many thoughts of gifts
special things for mom
to give her that special lift
weighing all the years past
Jesus, mom, and me
bad times are good times
with Jesus, mom, and me
it seems I was the toughest
new challenge every year
learn some independence
still needed to be near
the three other children
Jesus, mom, and me
a faithful Christian father
a happy family be
when times seem the toughest
it's Jesus, mom, and me
her advice and love like his
the strongest and always free
other siblings only wish
a life somewhat like mine
after trial and tribulations
a gift from him divine
everything goes smoothly
when I slow down and see
the focus has to be
Jesus, mom, and me

jump wings

we had all received our training
I guess it's eons ago now
yet every time I close my eyes
it seems the here and now
I remember the grass smell
as we worked through ground week
just starting to separate
the wild from the meek
the second week is gaining power
learning important skills
jumping the 34-foot tower
preview of upcoming thrills
week three is the kicker
it's where we get our power
first real glimpse of real airborne
two hundred fifty-foot free tower
week four is the kick in the pants
full of airborne delight
four jumps during the day
and hopefully one at night
five in all for each person
such amazing things
then we are awarded
our very special jump wings
for most of the jumpers
just pinned on their chest
but for truly airborne soldiers
they're punched into their chest

just a regular Veteran

usually sitting outside
look inside at the view
take a quick walk sometimes
just kinda passing through
good Lord then decided
here's whatchure' gonna do
come in and stay a while
then try to explain the view
help the average person
as far as our nation is wide
see the angels helpers
rescuing the Veterans hide
staff of pencil pushers
care not of flesh and blood
rather cause misery
tears in a family flood
late night or early morn
carrying the heavy load
nurses show what care is
turn the staff into the toad
true healing begins within hearts
caring and sharing people that work
not the pencil pushers
finding ways to be a different jerk
we find our greatest outlets
our empathy and consideration
by relating to one another
and a thankful reiteration
in our VA hospitals it's not
the rule makers who make it work
it's the smiling nurses who
smile at those jerks

just wanted

as a kid
all I really wanted
was to fit in
and not feel unwanted
when I grew up
I found my way
to hide these things
so deep away
became a man
what America wanted
best of the best
feelings blunted
always training
so far away
creating heartache
no one stayed
the soldier story
one and all
of how we live
and how we fall
I just wanted
to be the best
to never fail
at any test
best of the best
from the sky

hide the pain
and never cry
I just wanted
to be a man
to be accepted
in my own land
I just wanted
someone on the phone
to not feel
so all alone
I just wanted
every now and then
to just maybe
feel normal again?

knights

knights in white satin
was written for you and I
it was my favorite song
the day you caught my eye
we may have been young then
yet they knew our hearts
the moody blues nailed it
on every single part
so many letters I've written
only wishing to send
to maybe find you somewhere
bring this hollowness to an end
beauty has always been
with these eyes before
inside my heart you've been
and you'll be evermore
because I love you
yes, I've loved you
gazing at people
some hand in hand
just what I was going through
they'll never understand
because I love you
some tried to tell me
thoughts I'll never defend
just what the truth is

we'll see in the end
cause I love you
yes, I love you
oh, how I love you
knights in white satin
never reaching the end
letters I've written
always hoping to send
beauty has always been
with these eyes before
no one I've ever known
affected me like you before
cold hearted orb
that rules the night
heal this heart that it
may take flight

life notes

life is really
just bridges and roads
never expect more
than what you're owe
life is just
twists and turns
listen well to
never get burned
all of us happen
to make mistakes
see these admissions
can make us great
the only one
to be free of sin
died on a cross
to pay for our sins
align yourself
with hearts alike
saving the world
the greatest fight
swallow our pride
every single night
starting fresh
with the morning light
admit to ourselves
when we're wrong

an image to others
on how to be strong
grow a smile
every single day
help every single person
have a better day
when the Lord answers
one of your prayers
don't push it away
and live in despair
if changes are things
you secretly desire
don't shoot it down
and set it afire
life can sometimes be
fought all alone
it makes no sense
with modern telephones

lonely holidays

take care and treasure
each person in your life
so many issues come about
swifter than a knife
one day walking with us
the next a memory
every thought a source of pain
such an enduring misery
a mother passed away
long before her time
the family's strongest bond
slowly starts to unwind
a father's sudden loss
oh, so deeply felt
the family candle brightly burning
has slowly started to melt
the tragic and deeper loss
of any of our sons
the mighty pain and emptiness
that can never be undone
families will surely tell you
once a daughter has passed away
they're really walking with us
every single solitary day
it's not just the yuletide season
when the pain seems the worst

we seem to carry it with us
as though it was our curse
so please treat others gently
as you go about your day
you've no idea of the sorrow
on their shoulders as they pray
count yourself as fortunate
to never have had these burdens
put these things inside you mind
be gentle in acts and words then

Memorial Day

there are no words to tell you
what memorial day truly means
of my fellow Veterans
and those who lost their dreams
we worked so hard for the uniform
a right of passage you must win
train the body, mind, and spirit
a life of service to begin
we take our tour of duty
sometimes left alone
wives heading back to hometowns
feeling that no love was being shown
they felt they were really neglected
though they didn't really understand
that love, duty, honor, and respect
all go truly hand in hand
when we take the oath for service
when we sign that dotted line
we guarantee America
we'll protect her for all time
because it's not something we take lightly
through war after war after war
we stand up with each other side by side
don't mess with America any more
we back our brother/sister Veterans
help them get back on their feet

working with them, night and day
warms my heart, it's such a treat
for no greater people grace the earth
than those who gave their best
from the days of the musket
to the days of the bullet-proof vest
if you have a chance to read this
and had the American flag grace your shoulder
know that all of the country thanks you
you make every one of us bolder

mental health

mental health is
as most can say
a serious thing
affecting every day
a solitary struggle
to just stay alive
it actually affects
at least 1 in 5
complicated by
excessive pain
we feel our lives
going down a drain
everyday life
a struggle you see
how to continue
to still be me
it gets so bad
in heart and head
the pains too intense
why am I not dead
we've suffered things
others could not
we shuffle along
with spines in knots
the pain inside
hurting our heads

never really knowing
what words were said
some will go
by their own hand
we can only try
to understand
rock stars, writers
husbands and wives
suffering through
their entire lives
while doctors try
to understand
we can only hope
to make a stand
to scream in pain
oh help me please
this has driven me
to my knees
the pain inside
both head and heart
has ripped my entire
life apart
Lord, please oh please
I can't go on
help me make it
through one more dawn

military honors

military service
in one's life
helping our nation
through it's strife
uniform pressed
lines are straight
mind kept sharp
part of fate
respect became
a way of life
always on guard
both day and night
family, country
God above
taught about
eternal love
learned the roles
selfless sacrifice
faith in God
worth the price
freedom comes
as everyone knows
when haters dies
then love grows
the American Veteran
with an open heart

will always do
the hardest part
to pass beyond and
receive the Lord's grace
to pay the price
for the human race
each has given
more than received
comfort for
all the close bereaved
military honors
are not always
shiny, silver or brass
it's found in the hearts of
those left when they pass

my flag ii

these red and white stripes
flow softly do my legs
stars adorn my head
we were only following regs
we fell quietly from the sky
the mission in our heads
knowing whatever it took
one or the other would be dead
prayers for my family
escape with a gasping breath
prayers for my countrymen
with my dying breath
passengers on the flight fall silent
proud to respect one of their own
covered with the stars and stripes
on his final journey home
there never was a question
of who, or where, or why
as these stripes cover my body
and the stars are over my eyes

my grandpa

even as a young boy
just around about four
my grandpa had taught me
a gift forever more
as simple as bow knot
just learned to tie my shoes
I see him now in everything
that I say or do
bar, armchair, sofa
I was happy by his side
he talked to about life
in ways i'll here confide
he inspired my military spirit
taught me about being my best
with a flag like his on my casket
when finally laid to rest
he told me to always listen
twice then speak once
form your thoughts slowly
never once being the dunce

my spangled banner

please tell me, Lord
have I done well enough
to be buried beneath her stars
will her stripes
adorn my box
and hide my many scars
she gave me a reason
to stand and be proud
when so few would
I just hope you understand, Lord
times were hard
I did the best I could
it was for God and country
that I put my life on the line
and still will to this day
for those two put together
make Americans' lives just fine
and the price will always be paid

nights of roaring silence

imagine if you will a sunrise
with no light to see
nothing but the roaring silence
and the haunting memories
a simple type of childhood
running wild and being free
catching crayfish in the stream
and climbing every tree
teen years spent working
flipping burgers or bailing hay
clothes needed and car repairs
learned to pay our own way
nagging question in our head
how to repay for these rights
strap a parachute to my back
pick up a weapon and fight
terrorist killed our families
in attacks of mass scales
we must respond in kind
jump and never fail
patrolling hostile areas
hit a buried IED
ears won't stop ringing doc
why can't I see
shrapnel did it's damage
made the world go dark

damage to the ear drums
tore this soldier apart
never to witness another sunset
or hear a lover's whisper
life's everyday challenges
makes a grand plan seem sinister
how do you save the Veterans
who gave everything and more
treated badly by the hospital
ending up dead upon the floor
it's twenty-two a day now
soon it could be more
selling out our health
losing sight of the poor
they only take the smartest
to achieve a stronger unit
then kick us to the curb
when it's time to quit
and the Veterans that gave
all they could ever give
they remove their healthcare
give them no way to live
every single damaged Veteran
in one last act of defiance
sits alone in a dark room
listening to the roaring silence

oh, so long

she waited for him at the airport
it had been just too long
lonely days of silence
thinking of their favorite song
he went to serve his country
the best that he could be
standing up front for freedom
so that others could be free
high school sweethearts together
married by a justice of the peace
he shipped out to basic training
four years till his release
he then chose to go airborne
protectors of this great land
first to be activated
first to make a stand
with only a late-night phone call
tears and I love you
he promised to be careful
this is what he had to do
the letters were infrequent
she could only understand
honoring his commitment
serving in the burning sand
then came the notification
the chaplain at her door

her loving soldier husband
won't be coming home anymore
today she's waiting at the airport
one red rose in hand
she's waited oh so long
to stand next to her man
the flag draped box rolls slowly
funeral detail in dress greens
the tears flow so softly
it all seems just like a dream
she walks beside her husband
now covered by the flag
life was once a fairytale
it's now become a drag
she holds onto his picture
so happy and full of life
so very proud to serve
so proud of his loving wife
paid the ultimate price for freedom
for those he didn't even know
such a loving caring husband
that's why she loved him so
as taps plays in the distance
and all the mourner's cry
some may soon forget
she will never really say goodbye

on my back

on my back
atop this gurney
five thousand miles
started this journey
shrapnel wounds
roadside bomb
so much of life
tragically gone
IED
explosive device
my body's cold
feels like ice
what I'd give
for a call from home
just a few minutes
on the telephone
to share with loved ones
so far away
share the love
I wanted to say
my thoughts of you
mom and dad
brothers and sis
please don't be sad
I felt it best
to go this way

in this land
so far away
no alcohol
no drugs
no gangbanging
inner-city thugs
in a field hospital
in Afghanistan
is where we were
lying in the sand
messed up mangled
not all intact
not enough left
to fit in a sack
but one thing left
this American heart
no ISIS bomb
dare tear apart
please don't be sad mom and dad
you were the best
a guy ever had
my body destroyed
my spirit soars
welcoming new soldiers
at heaven's doors
so, my family
until all wars end
I'm safe in heaven
and my love I send

our military codes

any with a military past
each of us truly knows
meanings of certain words
spoken in military code
doesn't take a genius to see
when it could possibly be
o'dark thirty can always be
one thirty to five thirty
ever since time has stood
since the days of the druids
every single soldier has searched
for a gallon of blinker fluid
every single airborne trooper
since 1940 without cease
has spent at least a day
searching for riser grease
people look at us funny
maybe a little agitation
when we find the humor in
looking for a tank washing station
every soldier and his brother
whether you believe it or not
just missed expert marksman
by one or maybe two shots
every other soldier says
he feels cheated out of his shot

his recruiter screwed him out of
the jump wings they never got
every single soldier knows
we work when the oils glow
while others are quietly sleeping
at work our brain is keeping
as the holidays are drawing near
for soldiers of every kind
a sense of joy and happiness
for each I hope you find
a merry Christmas to each is sent
Marine, Navy, Air Force, Army
happiest wishes to all is sent
without sounding a little too smarmy
happiest of holidays to all

part of me is gone

there's another world inside of us
that you may never see
the battle-ready killers
we were trained to be
they're secrets behind these eye
we can never let you find
things that just might
make you change your mind
or maybe we're just blind
so, hold us when we're here
love us when we're gone
when we can't be there
you know we are aware
you know our hearts will bleed
with every act or deed
no matter what goes down
we always stand our ground
it's for the better good
our families understood
when we signed that line
we said that we'd be fine
even in the line of fire
and if we should expire
our families understand
that we're helping our fellow man
so, love us when we're gone

passage of time

like the earlier days
of rock and roll
growing older will
definitely take its toll
Jefferson starship
a song called Jane
cheap trick played
I will be the flame
all the love
that's come and gone
facing each new
rising dawn
no matter ages
clouds or sun
each day holds
love and fun
don't be fooled
by birthday candles
there's nothing in life
you can't handle
a happy smile
a hand to hold
the greatest joys
of growing old
to find someone
to share these days
are the gifts
we can't give away

picture of perfection

it's amazing what happens
we see in retrospection
what the Lord decides
looking in reflections
simple souls crossing paths
along life's many ways
suddenly bringing sunshine
on some of the darkest days
certain people bring friendships
far beyond the norm
touching your heart
weathering each storm
just the sights
of their picture bring
thoughts of sunshine
and warmest spring
the world becomes
a happier place
with that beautiful smile
on your beautiful face
a graceful heart
sets this world free
by touching hearts
like the one in me
no matter the speed
the world's growing pace
it'll always blossom
just seeing your face

real friends

every day
a Veteran knows
the tides of life
and where it flows
marine, airman
navy, soldier
friendship makes
each one bolder
life's events
big or small
so much better
shared by all
military mindset
life in order
different branches
all supporter
life much easier
as a clan
awake all hours
a helping hand
electronic media
each report
regardless of hours
provide support
cup of coffee
bag of food

stopping by
in the neighborhood
no greater love
like family be
all my Veteran friends
so special to me
lift my heart
every single day
light my path
the Lord's way
no greater trust
I've ever known
each Veteran friend an angel shown

selfless sacrifice

there comes a time
in each person's life
to take on a mate
husband or wife
then life becomes
bigger than ourselves
childhood things
put up on shelves
more earthly endeavors
come to mind
what's the legacy
we'll leave behind
peace on earth
good will to men
or terrorist acts
committed with pen
we take to heart
our place in history
good or bad
is no real mystery
some will take
a passive stand
while brazen others
draw lines in the sand
sign their name
on dotted lines

telling their families
that they'll be fine
letters from home
their daily bread
every line of ink
like each word said
an occasional call
to those they love
or virtual visit
through satellites above
some may get
no letters or calls
no reassuring words
before they fall
no thank you given
for their service that day
as that last simple breath
slowly drifts away
a simple thanks
would have been nice
for such a
selfless sacrifice
so many chances
to walk away
but every one
chose to stay

serenity

serenity is
as most
will see
another
stage of
serendipity
happiness found
every day
sharing friends
calming
comfort
without end
common ground
in just
little things
amazing what
each new
day will bring
comfort in
silence is
it's own reward
God's greatest
gift found
moving forward
encouraging
learning
equal support
healing hearts
soothing souls
heart's resort

service animals

service animals are
quite different you see
will sacrifice themselves
like the Lord upon that tree
they each only know
what happens in their heart
how they fret and worry
every second they are apart
they understand the needs
of urgent medical issues
or how to retrieve
a roll of bathroom tissue
most cannot fathom
the depth of their dedication
giving up their own lives
without a moment's hesitation
for any public power
to take one of these lives
should be a criminal case
for all these acts deprive
dog is God in reverse
here to teach us all
unconditional love and support
the happiness in a simple ball
so oft we see this
cops shooting service animals

time to really question
who's doing the service
and who's being an animal

This is dedicated to the millions of service animals that aid our Veterans and civilians alike to make life easier and more enjoyable. and it's a wake-up to the number of times that these animals are mistreated and/or killed because their human is having a crisis and the authorities don't want to handle things properly where the service animals will be treated with the same respect as police dogs. service animals should be given the same rights as Veterans or other public servants. thank God for dogs.

service connected

it once was
when service connected
what people saw
was easily connected
times have changed
duty more intense
serving in our
nations' defense
the things they've seen
and had to do
kept tucked away
from civilian view
nightmares raging
and still they go
back to fight enemies
most will never know
when asked about why
they chose to return
they always say
can't let my friends burn
I'm needed there
much more than here
the look in their eyes
says that it's sincere
service connection is
now you may see
our lives on the line
for eternity

set free

the darkest night
the softest plea
is a giant wave
on the human sea
crimes of souls
crying in pain
too many lost
nothing to gain
throughout history
again and again
controlling force
forced to bend
change the future
every single day
learn forgiveness
along the way
the Jews forgave
the evil king
to find the enlightenment
freedom brings
the jews forgave
the Nazi regime
to find happiness
and start to dream
the hardest part
we all know

is saying stop
and letting go
we all must choose
the ways to live
we each must first
learn to forgive
each tomorrow
a bright new day
the way we act
will lead the way
lead with purpose
strength of heart
share God's blessings
and do your part

set your pack down

being labeled a Veteran
puts you in a different class
a mean green son-of-a-bitch
the world can kiss our ass
but being trained to bottle up
to swallow our pain and go
our lives are so different
than the regular world would know
to ask for help is a weakness
or so our brains tell us is so
you are an American soldier
pick up your pack and go
it's time to set your pack down
time to heal your head and heart
time to put you back together
time for a fresh new start
you've served your country well now
on land, sea, and air
you can set your pack down
and trust us with your care

she or he died

all too soon
our lives may end
leaving family
and many a friend
tears will fall
words then shared
speaking of how
life's unfair
they may have
flown the skies
served up
burgers and fries
saved lives
first responder
sailed off into
the wild blue yonder
used their hands
built our towns
dressing up
being clowns
many lives
things to do
all as different
as me or you
make a difference
efforts applied
volunteering works
change never denied
but in the end
after all have cried
headstone still says
she or he died

sisters and brothers

sisters and brothers
each in arms
together we faced
the worst of harm
from 1776
until this morning
we left our families
sadly mourning
trust so deep
so, few know
nowhere on earth
we wouldn't go
in the air
sandy beaches
jungles and deserts
farthest reaches
standing up
because we must
only with those
we truly trust
took the oath
the dotted line
said our prayers
hoping we'd be fine
some took ships
others planes

their motives no one
need explains
for family
for country
for husband
for wife
for love
for liberty
and freedom
from strife
to grow
to change
to share
to care
to help
the world
to become
more aware
the human plight
to make it work
fewer people
being a jerk
every soldier
every girl or guy
in camouflage
ready to die

the whole world
their saving grace
willing to save
the human race
taking a stand
for what is right
passing gently
into the night
they carried out
their greatest call
no one there
to see them fall
brothers and sisters
American pride
changed the world
turned the tide

skeletons and shadows

in every life lies
skeletons and shadows
bits and pieces
they don't want you to know
little bits of information
words they can never say
hateful things yesterday
that never will go away
parts of someone else
or so it would seem
waking up at night
in an almost silent scream
no one to have
no one to hold
like we're formed
from a broken mold
every single night
in the witching hour
our breath quickens
then grows sour
yesterday comes a' creeping
like a nightly villain
full of dread and worry
each of us is filling
some past events
no one should know

so late at night
with an evil glow
in each closet
or hidden space
tucked away from view
no one to deface
every person has
their own piece they know
a different name to
their skeletons and shadows

someone

the soldier's name
was Someone
we see every day
they served in honor
at different times
for so little pay
the soldier's name
was Someone
service done with pride
things go wrong
it seemed like hell
trying to remain strong
the soldier's name
was Someone
man or woman with a soul
giving all to country
damage to a fragile mind
finally takes its toll
the soldier's name
was Someone
trying hard to find care
just closed doors
just wait some more
it's more than just unfair
the soldiers name
was Someone

lost to addiction or suicide
feeling alone and empty
no real remedies
no place to run and hide
that soldier's name
was Someone
near and dear to your heart
a better system is needed
to save these lives
create a brand new start
that soldier's name
is Someone
you know and love so dear
a better answer is needed
to save 22 a day
a message oh, so clear

stand up

at different times in life
when you feel like giving up
take life by the horns
find your legs and stand up
when life hands you pits
instead of the bowl of cherries
just straighten up your spine
life can soon be merry
don't be afraid to stand up
when others put you down
soon the proof will show
who's left looking like a clown
don't take it lying down
till the day you're fed up
make sure your back is turned
now finally just stand up
no more dirty insults
or double standards
fire back with both cannons
no fear of the lanyards

stranger

take a look
and tell me
what you see
just another
man or woman
normal so you see
hidden behind
each one's eyes
pain held deep
haunting nights
haunted days
interrupting sleep
checkout line
holding signs
struggles to get
by fights addiction
with own restrictions
never see them cry
the Veteran unseen
though rarely mean
even the trained ranger
look just beyond
the light comes on
you know who is
this stranger
they are a Veteran

struck by lightning

have you ever
had an event
that maybe just maybe
was heaven sent
like every word
a description made
an answer to
every prayer prayed
the perfect face
infectious smile
has stayed with me
for quite a while
a special heart
to help others
all God's children
sisters and brothers
was our meeting
supposed to be just like
or possibly more
with a lightning strike
it's totally cool
and all depends
where this goes
as more than friends

suffer the children

word was once
suffer the child
spare the rod
spoil the child
small and fragile
angels at heart
parents tearing
these lives apart
abuse is silent
left alone
another child
and a telephone
no connections
like plastic dolls
do these people
care at all
leaving scars
no one sees
silent screaming
echo off the trees
the reality is
all too true
it actually happening
in front of you
listen to what
the children say

make the pain
go away
left alone
like no one cares
we all see
this isn't fair
be the one
to hold them tight
be there for them
every single night

tears for the fallen

the American flag snaps briskly
in the cold and driving rain
flown at half-mast again
more Americans slain
yesterday it was our soldiers
our Veterans coming home
then hate groups took over
and things have overblown
blaming for a virus
blaming for nothing at all
targeting school children
walking down the hall
judging by colors
by clothing each may wear
by the way each speaks
the colors of our hair
too many tears are falling
hatred on the loose
hateful messages sent
like a mask or a noose
the American flag is dripping
from all the fallen tears
never so much hatefulness
in over a hundred years
we must calm the violence
put our hand on America's heart
stand with one another instead of far apart

tears

when we all were younger
tears had served us well
constantly springing forth
from a bottomless well
shedding tears at weddings
funerals bring waterworks too
the deepest type of heartache
whenever it happens to you
thoughts of missing loved ones
things that might have been
all of the things you tried to say
just cast out to the wind
all of the special moments
tiny lessons you've learned
touching your heart each day
with just the faintest burn
simple soup and sandwiches
footprints in the sand
being just four years old
with shoestrings in your hand
ice cream in the summer
watching airplanes come and go
the future creeping up on us
adults before we know
don't be afraid to shed
a tear or two, it's okay
it helps to cleanse the soul
and show your heart the way

the American soldier

the American soldier
half man half machine
always on call
even as you dream
even after leaving service
and all they've been through
putting on their boots
for the things they must do
the things they've seen
and done for you
just for the freedoms
that keep this country true
in the land of freedom
from harm and from fear
there is always one of us
bravely standing near
freedom to be the person
you really want to be
from the highest mountain
and sea to shining sea
respect for our country
for man, woman, and child
protection for each person
even the meek and mild
our obligation has no limits
until our days have gone

we pledge our allegiance
with each rising dawn
when wolves prowl in darkness
on unsuspecting sheep
we are there to protect you
and put the wolves to sleep
the American soldier
trained to do what's right
trained to be ready
to stand alone and fight
the American soldier
the best fighting machine
world's worst nightmare
great American dream

stages

life consists
of different times
different stages
people we meet
stages of life
different pages
each day
builds upon
the next
different people
different reasons
all intersect
each person
a teacher
good or bad
preparing us
for tomorrow
happy or sad
every situation
a lesson
in humility
how to care
how to share
the tranquility
learn how
to listen well

to more than words
sometimes it's all
about things
you never heard
all of life
people and stages
different times
different ages
same lessons
at different stages

to each it's own mile

each life is lived vicariously
as each feels free to do
myriad choices to be made
as longer each day always grew
each knows the paths they've followed
the events filling each day
the truth behind their motives
regardless of what others say
some make big efforts of showing
an appearance here or there
never flexing real muscles
just pretending to care
others make time for people
give time for those in need
helping to teach spirituality
by planting positive seeds
some may have lived in bitterness
a past that would not let go
pain and despair in tatters
every place that they go
others have survived the damage
caused by others twisted minds
they finally reached the revelation
that it was in the grand design
no matter the length of the journey
life is a long trip to endure

there should be trust around you
with love and light in store
while we each wish for perfection
when we need to sit for just a while
accepting life's abstractions
and finding our reasons to smile
while some may have been ministers
other firemen, soldiers, or wives
teachers, doctors, nurses
all different kinds of lives
please don't judge others harshly
forcing them into denial
feel free to live your life
and to each its own mile

thirteen

in a massive crowd
in a faraway place
evil rose up
and showed its face
thirteen warriors
lost in vain
13 flags and memories
all that remain
millions of flags
hanging half staff
who's to blame
for the bombing gaff
a nation's tears
continue to fall
as groups of Veterans
visit the wall
so many good souls
lost to war
our society cannot take
this any more
shell shocked victims
women and men
battle each day
with the poison pen
sights of hell
behind closed eyes

yet called monsters
and clearly despised
instead of grief
and living hell
try things instead
to make them well
we need to stop
this deadly flow
save every soldier
each of us knows
no more caskets
draped with flags
no more taps
or half-staff flag
find new ways
to handle terror abroad
killing innocents
and committing fraud
thirteen souls
next time more
show these heathens
how we even the score
as a general
to the enemy said
who'll raise your kids
when you're dead

souls lament

blood coursing
like cars
on a highway
each has
drops of love
the skies
above today
dark portent
mottled grey
but still
life's okay
multi-color
vehicles
buzzing past
each different
from the last
yet beauty
by contrast
each color
on display
as the clouds
break away
few words
truly say
the world
yesterday

and or
today
was changed
in every way
every word
proven true
every sky
azure blue
all because of
just you

the picture window

on a hill next to a meadow
a house built on a concrete slab
never been touched by paint or brush
it's existence bordering on drab
the door hangs off its hinges
the handle coated with rust
as if to resist entrance
enter, only if you must
many an antique cobweb
dust and dirt high and low
tobacco stain rainbow swirls
never a light bulbs glow
numerous glassless panes hanging
nailed shut to restrict only large prey
who knows what nocturnal visitors
may enter and decide to stay
as the sun falls past the crown
like fingers reaching through hair
the specter appears in the window
in the front, one large square
no one knows the history
of how she came to be
wavering swirling reflection
next to a plastic tree
her passing came at yuletide
she wasn't quite twenty three

riding her mighty white dragon
in front of the Christmas tree
her family had grown distant
two tours of bitterness
no ringing cell phones
no one there to witness
benefits used to rent
a shack on a concrete slab
leaving from the hospital
in an antique yellow cab
no staffers doing follow ups
the nightmares clinging tight
gotta ride that dragon
keep the monsters outta sight
military sexual trauma
post-traumatic stress disorder
she became more distant
as her time grew shorter
family just grew indifferent
uncaring and complacent too
said that she was faking it
that those things could not be true
she just silently walked away
the world had gone wrong
now it feels it is a place
where she no more belongs

finding a gentle out
go soft into that gentle night
no more painful suffering
an end to this blight
as the sun begins to rise
the film shares a radiance
she finds her heavenly release
gaining the Lord's guidance
the lady in the picture window
next to the plastic tree
has been received by God
and spirit finally set free

true soldiers in green

as I walk the halls of the hospital
during the day, late at night
so many men and women
bearing the scars from their fights
some just barely drinking age
others octogenarians
all have answered that special call
all are American Veterans
so many have been injured
rescuing their brothers in arms
never thinking of themselves
suffering bodily harm
so many suffer injuries
training for the inevitable call
pushing themselves far beyond
to be the best of all
some injuries are not visible
they suffer in silence and dread
not ever knowing the reality
of the anguish inside their head
it takes a certain individual
to volunteer to stand in front of a bullet
to stand up to any enemy with a gun
and say, "go ahead asshole, pull it"
as American soldiers we know
we have the good Lord on our side

we treat the world as we would want
and force all evil to hide
I pray each night for each of you
may God give you comfort from above
may he calm and ease your aching heart
and fill your days with love
for each of you have given
the best part of who you are
without your dedication
we'd never have come this far
thanks for your service

the earth's been hushed

the earth's been hushed
brought to its knees
only things moving
are the birds and the bees
every single person is
sheltered in place
rarely going out
respecting others' space
suddenly each river
has gone back in time
people are viewing mountains
because of clearer skies
life seems so much better
a stronger sense of love
like people realize
there is a God above
he would never release
a virus killing thousands
yet receiving each soul
into his loving hands
could this be a message
from God high above
stop with all the hatred
treat each other with love
treat this garden I created
as a wonderland for all

don't think yourself so large
when you're really quite small
this was meant to be a garden
never a waste land to be
a beautiful place for my children
for all of eternity
here is his message simply
to each and every nation
to every religious following
regardless of denomination
the earth has been hushed
so that each of you can see
what a beautiful place this is
for all of us to be
if we don't grasp this chance
to rescue one another
our final chance is coming soon
and we will not have another
that is why
the earth has been hushed

the sharpest blade

it is easy to pass judgements
call someone out for nothing at all
do you even understand
what precipitated their fall
for some it was abandon
no one to hold on to their heart
society keeps pushing them away
never allowing to be part
for others it is called abuse
some physical, sexual, or mental
just the damage it has caused
cannot be called incremental
for some it is misunderstanding
refusing to see the miles each person tread
perhaps calling them a slob
for never making their bed
or judging the way they do things
can you walk a mile in their shoes
have you removed your shoes
to feel the pain they feel when they do
maybe they are suffering a setback
an obstacle that most do not
but face life every day
with a smile that can't be bought
so, before you belittle or insult
cutdown or verbally degrade
think how you would feel
to be cut by the sharpest blade

whispers in the silence

whispers in the silence
when you're all alone
so late at night
you hear whispers in the silence
you find your strength
to do what is right
and claim your independence
when you feel your fears
rising up at night
when you feel so all alone
you know you can reach out
make things alright
as close as the telephone
I never claimed
to be a sage
a man with ultimate vision
I only offer
my caring heart
a wish of no imposition
so always know
I will be here
if you should ever need
I will listen with
an open heart
if that's what you need

tiny pieces

it seems to each of us
that the world has been shattered
everything that meant anything
broken up and been scattered
when we lose our mom's
it shocks our brains
what's going on?
am I insane?
everything hurts
nothing matters
life is left
all in tatters
and yet you will find
you are no longer blind
your heart will heal
you're on an even keel
she knew one day
you would feel this way
she taught you love
with a velvet glove
she taught you hope
so that you could cope
she opened your eyes
to the morning skies
so, the pain in your heart
would slowly fall apart

it would help you sneeze
with every morning breeze
she taught you to love
from the Lord above
her physical form
may have passed from this plane
but the tiny pieces she gave you
will always remain
for the gifts she gave you
as a mom always will
you have passed to your child
to carry on still
the tiny pieces of our mom's love
are life's hardest lessons from God above
tempered and softened, held in our hand
helping us through it, as only mom can

the pack on your back

we each travel each day
with our pack on our back
in it are the treasures
starting many years back
something's may resemble pebbles
picked up in a stream
tiny little fragments
of long forgotten dreams
tiny slips of paper
a word or two written down
a few faded tickets
from the circus with the clowns
several tiny pictures
of a child since grown old
a family out making snowmen
ignoring the bitter cold
a single letter from a father
wondering why he's alone
no letters in the mailbox
no ringing telephone
he tried his best to give to others
offered to help in anyway
she had just decided
was best to turn him away
inside a faded bible
filled with markers here and there
he spent his days with others
showing them how best to share

we each carry our own pack
filled with treasures from our lives
memories of our children
good times with our wives
each may have had a husband
and heavy loads to bear
they always knew you were there
they always knew that you cared
the contents of our own packs
are treasures only we would know
until the Lord calls us
and says it's time to go
the Lord understands our treasures
sees the values we hold in our hearts
says "it's time to rest now"
"you've surely done your part!"
he understands the heartache
the loneliness and despair
the many nights of feeling
like no one was really there
that's why he gives us insight
into how others feel
because our packs are heavier
we already know the deal
so, collect your memories slowly
be careful, don't lose track
it's as original as you
it is only your pack

thankful

a dollar seventy-six
left in the bank
yet there's food
the Lord to thank
a biscuit here
with gravy smothered
a daily prayer
the Lord has me covered
planted some veggies
asking the Lord's help
planning to share
on many a shelf
plenty of sunshine
a bucket of rain
planting some sunflowers
my hopes will explain
the Lord's loving spirit
and kindness and love
things we can share
from high up above
giving us faith
from day to day
helping us to find
the right things to say
like each precious vegetable
each garden provides
the good Lord restores
our hearts deep inside

treat us gently

seasons are slowly changing
everyone's heading to the mall
searching for that perfect gift
for favorite people big and small
most will gather for dinners
ham, or lamb, or tofurkey
we just try to accept each other
even when a little bit quirky
we mustn't forget the silent ones
that tend to sit back and suffer
or that tend to act out a little
just to create a buffer
we just want you to love us
to accept us as we are
if you knew what we've been thru
you'd see we've come so far
some days are perfectly normal
when we get up out of our beds
other day's there's this screaming
going on, inside our heads
it's like the winds of winter
howling across a ships bow
we just wish we could find a way
to stop the screaming now
for others it's a whisper
keeping them up at night

cowered in the corner
giving them such fright
for others yet still the battle
seems to go on without end
constantly haunted by memories
of how they lost their friend
we can only tell you
of the pain in our heads and hearts
and how all of these feelings
make us feel far apart
so, as you enjoy your dinner
as you bow your heads for grace
please be understanding when
we ask for a little space

springtime springs

barren trunk and branches
waving to magical music
that only they can hear
cars, trucks, bicycle, ATV
everyone out and busy
ready for spring to be
boys and girls holding hands
hoping that this love
will someday truly be
sunny days and chilly nights
trees shaking buds to their tips
as winter finally releases its grip
senior citizens recovering their health
making good use of their wealth
walking away from the tomb
New Year's resolution kept still
even lower medical bill
less time in the living room
time at work, texts, faxes
April looming, paying taxes
dream of a refund
soon the heat of June
stay indoors once more
or bake like a giant bun
as a final call
the leaves will fall
and be cold outside once more

while everyone sleeps

while everyone sleeps
and rests their brain
some of us deal with
excruciating pain
expensive pillows
and meds galore
acupuncture needles
till we're stuck to the floor
the flag whipping
outside of my door
I sometimes think
I can't take much more
it feels as though
my head will explode
scoop up the remnants
flush them down the commode
bad jumps in service
they know the facts
they would rather pretend
this is an act
then they wonder
why are things this way
will I end being
one of the 22 today
service to country
selfless by every deed
refusing to give us
what we really need

the old man in the fall

the old man sits lonely
riding the early train
another fall afternoon
how many still remain
leaves are changing colors
goldenrod and bright yellow
dappled with morning sunshine
riders all quiet and mellow
leaves of crimson and burgundy
as every day grows colder
so many people being much nicer
as each of us grows older
as the train silently slides
past tree of all shapes and heights
they flash their amazing colors
in the headlights late at night
while we are young
we don't pay much attention
never thinking about those colors
and things like memory retention
what if you never got to see
the changing leaves of a tree
how much different would life be
just because of colorful trees
the old man is us you see
we are each the leaves you see
our planet is those trees
see how beautiful things can be

the silent march II

each soldier walks
that silent march
heaven bound
to the golden arch
the pearly gates
silently await
this soldier's stride
clean and straight
as each tear falls
as you pass
your work is finished
and done with class
uniform straight
brass reflections
ready for
the Lord's inspection
serving country, family
community, so well
your story is one
we know so well
lifetime spent
protecting the herd
living life
in God's word
the sacrifices made
for all to see
a greater country
and people freed

when we sign

in years long past
we had the draft
to fill the needs
of the military
forward then
fifty years plus ten
the military is
completely voluntary
you will see
we've kept America free
without thought
of the price
every soldier's heart
sets them apart
because they
offer the same advice
we'd serve again
knowing our end
that we may
make the ultimate sacrifice
when America calls
we answered that call
without hesitation
that it may take our life
we know when we sign
that may be the last time
we get to spend
time with our family
to keep America free
we give our life gladly
to save every
earthly family

every soldier's battle

every soldier's battle
spanning all of time
one of abandonment
of being left behind
every battle fought
since the start of time
forming lasting bonds
in the worst of times
whether bow and arrow
hatchet or M-16
every single soldier knows
things you've never seen
they never asked for pity
you'll never hear them scream
hidden very deep inside
a very terrifying dream
it may manifest as depression
or another type of madness
every soldier carries with them
a high degree of sadness
a fascination with giving
life for fellow man
best to give your life
making a final stand
every single American
anyone worldwide

the soldier will protect
with their greatest pride
the soldier's battle being
a single matter of life
regardless of those left behind
his family or his wife
every single soldier
battles to see each day
as a reason to continue
as a reason to stay

to what a man's soul

to what a man's soul
shall be gained
as he struggles through life
being completely vain
to what a man's soul
shall be included
if he spends his days
with drink diluted
to what a man's soul
shall he inherit
if he accepts all
of his challengers' demerits
to what a man's soul
has he to gain
when every word
he must explain
to what a man's heart
can he finally see
why his days are different
than you or me
to what a man's heart
as he suffers through hell
memories of bullets
rockets, and mortar shells
to what a man's heart
shall come to pass

as his body lies
underneath the grass
how he gave his man's heart
to those he loved
passing on the spirit
of the Lord above
he played his man's part
in the war called hell
carrying rockets
bullets, and mortar shells
to what a man's soul
shall come to be
standing up for
both you and me
no tears have fallen
never touched the grass
when this fallen hero
so silently passed
the raindrops are tears
from angels on high
bidding this hero
his final goodbye

the Veteran effect

just sitting here watching
Veterans scurrying to and fro
just grabbed a bite to eat
appointments to which to go
some zoom by on scooters
some are pushed in chairs
yet not a single one
whines "life is so unfair!"
some sit with families
others with wives or alone
yet in this sacred place
they're never really alone
some just grab some pizza
full meal with Diet Coke $^{(R)}$
plenty of staff around
in case someone should choke
it is still a meeting place
a place to make new friends
this is part of the healing
as your head and body mends
it is called the VA hospital
the place that helps us heal
the miracles that happen here
yes, they are very real
we rise to aid each other
a hand, a word, a smile
just taking the time
to sit with you awhile
so, you see this is greater
than any hospital in the land
from one Vet to another
we give each other a helping hand

the guard

no one but the soldier
understands the silence of the night
pulling all night sentry
until the morning light
whether in America
or on some foreign land
we stand ready thru the night
to make a kickass stand
we know who we are facing
what and how it must be done
walk our post or Fireline
armed with rifle or shot gun
the password may be simple
one no one would ever guess
heaven help if you forget your half
get a bullet in your ass
the guard is there for protection
for each man woman machine
keeping America ready
for heaven hell and between
as I walk my post slowly
I listen and notice what most would not
seeing even in the dark
among the many things we're taught
we started as ordinary people
wearing t-shirts shoes and jeans

a slow transformation
to a lean mean fighting machine
we love to guard America
and the freedoms of good people everywhere
and if you mess with any of them
we take it as the worst kind of dare
so, rest your head easy tonight
we stand ready for your call
we stand guard and walk our post
and never falter or fall
we take it as a calling
that all people should follow their heart
so, we provided the Lords protection
Christian values to impart
for the Lord would have us help others
escape from tyranny and oppression
they taught us how to eliminate our foes
in fairly rapid succession
so, rest easy we stand guard
ready to answer the call of any in need
we have trained long and hard
just to make sure we succeed

the dotted line

every man or woman
who's reached that state of mind
knows what it takes
to sign that dotted line
it never means that I am better
or that I care more at all
it means that I want to stand up
for American's big and small
it means that each life is precious
more valuable than gold
that America is my country
and I am one of the bold
I'll gladly miss my family
spend years out in the sand
battling the terrorist
that threaten our great land
when my tour is over
with DD214 in hand
I will silently watch over
the people close at hand
for when I signed
on that dotted line
I swore on my life
till the end of my time

the lone veteran

an innocent teen
just starting life
the dotted line
a brand-new wife
soon to learn
no days of bliss
marriage can be
truly hit or miss
the young wife
goes running home
never answering
the telephone
the lone veteran
serving his best
his heart and mind
can find no rest
daily dreams
of children's lives
stolen from them
by bitter wives
no simple walks
no Christmas trees
no bicycles
or skinned up knees
no tears to dry
no hearts to share

how can life
be so unfair
he chose to serve
he gave his all
she made his heart
take the fall
for forty years
his child stays lost
isn't that really
too high a cost
heart and life
put on the line
this is not
the Lord's design
open your heart
this father's day
make a connection
there's always a way

www.ingramcontent.com/pod-product-compliance
Lightning Source LLC
Chambersburg PA
CBHW060526080526
44586CB00012B/627